**From Panic to Power with Public Speaking...Drop the Mic!**

*A Neurobiologist's Personal Perspective*

By Sabrina Kay Segal, Ph.D.

Copyright © 2016 Sabrina K. Segal, Ph.D.

All Rights Reserved


Please direct all inquiries to: dr.sabrinasegal@gmail.com

Front Cover Photography by Eusebio Viajar

Back Cover Photography by Casey Dee

"In this book, Dr. Segal explains why public speaking is such a common and often debilitating fear, how to overcome that fear, and turn it into a strength. Using analogies from everyday life to explain the most up to date scientific research, Dr. Segal provides revoluntionary and simple solutions for how to become a confident, engaging public speaker. Whether you want to simply communicate more clearly in your social circles or you are speaking to an audience of thousands, this book gives you the tools to do it in a way that will 'wow' your audience and leave you feeling like a pro."


Blue Haught - Hospital Director of Surgical Services

## ACKNOWLEDGMENTS

Special thank you to my family for their feedback throughout the writing of this book, and for being part of my personal journey on which this book is written. To my amazing parents, Karen and Murray Segal for being the most incredible role models throughout my life and for always believing in me. To my sisters, Laurie Garner and Jordana Kopin for their continuous love and support.

To Chalene Johnson, my mentor in fitness and in personal development. There are no words to describe how much you've influenced my life. You have taught me how to find my strength, both in fitness and in character. I have pushed myself through challenges and accomplished dreams I'd never even imagined before I met you, because you inspire me to reach my full potential in every aspect of my life. Thank you for being such a light in this world.

To Bo Eason for inspiring me with your presence, with your story, and with your words. I have never met someone who moved me so much in such a short amount of time. Thank you for being part of my story in this book. In the same way that Jerry Rice inspired you with his "generosity", you have inspired me with yours.

Finally, thank you to all my friends, followers, and supporters. I believe that within each of us lies a love for science at a young age. I hope to re-awaken that love in as many people as possible. Thank you for helping me in this endeavor.

"Do not let your fire go out, spark by irreplaceable spark, in the hopeless swamps of the approximate, the not-quite, the not-yet, the not-at-all....The world you desired can be won. It exists, it is real, it is possible, it is yours."--Ayn Rand

**TABLE OF CONTENTS**

Chapter 1: A Fear Greater than Death—Calm Down, It's Just Adrenaline

Chapter 2: Guilty Culprit #1—The Spotlight

Chapter 3: Science-Based Tips to Combat Fear

Chapter 4: Science-Based Tips to Build Power

Chapter 5: Guilty Culprit #2—Drawing a Blank in Front of Everyone

Chapter 6: Guilty Culprit #3—Imposter Syndrome

Chapter 7: Guilty Culprit #4—Perfectionism at its Best

Chapter 8: Overcoming the Panic Recap

Chapter 9: Let Your Power Build

Chapter 10: Finding Your Power in Going Live

Chapter 11: The Power of Your Voice: Podcasts, Webinars, and Online Courses

Chapter 12: The Power in You

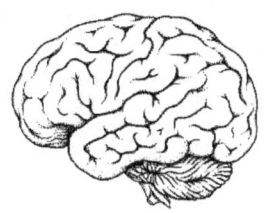

"According to most studies, people's number one fear is public speaking. Number two is death. Death is number two! Does that sound right? That means to the average person, if you go to a funeral, you're better off in the casket than doing the eulogy." —Jerry Seinfeld

## INTRODUCTION

Thump-thump. Thump-thump. Thump-thump. Your heart is pounding out of your chest...It's beating so loudly that surely they can hear it across the room. It drowns out every single sound, as you feel all of their eyes on you. Your breathing quickens, you start to sweat, and suddenly you can't stop fidgeting. You stare at your hands, and realize how nervous your jitters make you look, and then the panic sets in. With a timid nod you glance up at the crowd, all stoic faces just staring at you. Hands shaking, you reach for the microphone. You focus on it, trying to steady your breathing, hoping that somehow in the midst of this, you can remain calm and remember just what it is you are supposed to say. Your mouth is so dry that you're sure your voice will crack when you begin to speak, and everyone will laugh at you. You've never felt more powerless than in this moment. It's as if a terrifying dragon is blocking your path and every time you try to take a step forward to combat the fear, he breathes fire into your face.

Okay, so maybe you've never seen a dragon, but if I just described something you've experienced, you're not alone. In fact, you're one of millions of people around the world who share the same intense anxiety from public speaking. In today's world of multiple communication platforms—from formal presentations, to webinars, to podcasts, to live streaming, this skill has never been more essential to expanding the reach of your message. Overcoming the fear of public speaking is one of the key skills necessary to developing your career, building your brand, and growing your business.

My name is Sabrina. I'm a neurobiologist. I spent over twelve years obsessively studying the sexiest organ there is. No, get your mind out of the gutter! I'm talking about the human brain. It can be annoying, anxiety provoking, motivating, and inspiring. It can often be all of those things at once. Lucky for you, my research over the last twelve years has focused on the neurobiology of stress. But not in the ways that most people think of stress. Right now you might be picturing someone running from a bear. That's not the kind of stress I'm talking about. Or, you may be thinking of yourself in the wee hours of the night, drinking straight from the coffee pot and racing to meet a work deadline. That's not the kind of stress I'm talking about either. What many people don't realize is that stress can actually be a good thing—an actual *aid* to performance. Indeed, *if* you know how and when to use it, stress can be an incredibly powerful and helpful tool.

Whether you're 23 and preparing for a sales job, or 55 and thinking about taking your business to the next level, this book will be unlike anything you have picked up before. Maybe you are at the point where you're looking around and noticing all the people jumping on social media with live presentations. Maybe you want the reach, fame, and financial success that can come from mastering various public speaking platforms. But you think to yourself, "well that's just not in the cards for me," because the thought of just stepping on stage or hitting that Facebook "go live" button, makes you break into a sweat.

Well, my friend, you've come to the right place. See, your brain with all it's trickery and nuances, is one amazing organ. And the best part? You're in the driver's seat. So buckle up because I'm going to take you on a hilarious, profound, heart-breaking and awe-inspiring journey of how I took the panic I felt toward public speaking and turned it into a dragon slayer! So come on, we've got a bit of work do. And then my friend, it's time to smile, to feel that pride wash over you because you have a message that the world is waiting to hear. It's time to break through the silence, say what you want to say, and then...drop the microphone!

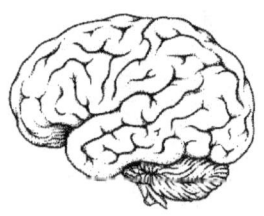

"There are only two kinds of speakers in the world: the nervous and the liars."—Mark Twain

## Chapter 1: A Fear Greater than Death—Calm Down, It's Just Adrenaline

I remember it like yesterday. I was about seven years old. We were playing outside and I ran into the street after a bright red ball. All of a sudden I heard my mother scream. I froze, paralyzed. I heard the tires screech and saw a car out of the corner of my eye. I felt something hit me in the back and suddenly I was stretched out on the sidewalk. I turned and looked up to find frantic, brown eyes that looked like mine staring back at me. My sister had run into the street after me, and pushed both of us out of the way of a car, within a split second.

I'm sure you've read the stories where people find the strength to lift a car off of someone or haul their neighbor out of a raging fire. It is in these moments that something powerful pumps through your veins. This, my friends is adrenaline.

Adrenaline is known as part of your "fight or flight" response. It also functions as a stress hormone. At a young age I became obsessed with this "superpower," this endogenous (made in the body) chemical elixir, that seemed to offer superhuman strength and lightening mobility. I had to understand what exactly had enabled my sister to push me out of the way in the nick of time. So you could actually say that it was an interest in the connection between adrenaline and the brain that led to my love for neurobiology.

You may have heard the term "stress hormone" thrown around as if there's like a zillion different types of them. Actually, there are only three types of stress hormones. Two of them are related as part of the system that activates your "fight or flight" response: adrenaline and noradrenaline.

Adrenaline gets its name from the organs—your adrenals—that release it when you encounter a threat. Adrenaline stays in the body to help you mobilize your muscles so you can get the hell out of a bad situation, or run across the street in time to save your sister.

Adrenaline does this while simultaneously signaling your brain to release its fraternal twin, noradrenaline. The two molecules differ by one methyl group (I promise that's the most chemistry I'll mention in this book). Despite the similarity in chemical structure, however, these two molecules have profoundly different roles.

In the body, noradrenaline, like adrenaline, functions as a hormone; but in the brain, noradrenaline is a powerful brain chemical (i.e., neurotransmitter), that can dramatically enhance memory (oh, also: in case you've seen the term "norepinephrine," that's the same thing as noradrenaline—but for the purposes of this book, I'll stick to using the term "noradrenaline.")

Your "fight or flight" response is initiated by a region in your brain called the hypothalamus. It signals the activation of your sympathetic nervous system (SNS). Think of your SNS as being your first emergency responder. Whenever you are exposed to a stressful situation this system is activated. The hypothalamus signals your adrenals to release the previously mentioned fraternal twins, adrenaline and noradrenaline.

Depending on the type, duration, and intensity of the threatening situation, your hypothalamus may decide to activate a second stress system. This system seems intimidating mostly because the name sounds like a Roman gladiator, but it's referred to as the hypothalamic-pituitary-adrenal or HPA axis. This pathway projects from the hypothalamus, to the pituitary gland, to the adrenal glands, and it's named after its anatomical roadmap. For simplicity sake we'll just call it by its acronym from now on, HPA. Think of your HPA system as your SNS calling in "back up" as your stress meter increases.

When your HPA system is activated, it releases the third stress hormone, cortisol. This is the hormone with the infamous reputation. He's the chemical on talk shows and commercials who's always blamed for belly fat, memory problems, and inflammation. While chronically elevated levels of this guy are not good for your brain or physical health in general, cortisol in short bouts can actually be enormously helpful for your memory.

Forget about adrenaline for the purpose of this book. We'll be spending a lot of time getting to know noradrenaline and cortisol, because they are precisely what provoke the paralyzing physiological responses that you may feel when you think of public speaking. Once you understand those responses, you'll become a master at controlling them, allowing them to work for you. This is how you will be able to convert panic to power—and to slay that dragon!

Now, let's turn to some of the most common reasons and triggers for the panic that you might feel toward public speaking. See if you can relate to any of these.

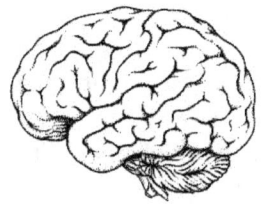

"My greatest fear is speaking in public. I'd rather not have everyone focus on me."

—Jennifer Love Hewitt

### Chapter 2: Guilty Culprit #1—The Spotlight

If I asked what scares you about speaking in public what would you say? If you're like most people, one of the first things to jump into your mind is the fear of being in the spotlight. You're afraid of being the focus as people are watching, evaluating, and critiquing every tiny aspect of how you look, sound, and move. And then of course, you are afraid of being evaluated on the content that you're delivering.

I know all too well the feeling of constantly being evaluated by others because I am an identical twin. All my life people would stare at us, and go back and forth trying to figure out what small differences there were so that they could tell us apart. It felt like being under a magnifying glass, the way people would stare. You can imagine how fast that gets old, when you have no makeup on, and you think there might be something in your teeth. We actually grew up near Hollywood, and were constantly in the spotlight. My mother had found us an agent by the time we were born, and we were in our first movie when we were 19 days old.

I hated going on interviews and trying out for different parts. But being on the movie sets was fun, and by the time we were ten years old, my twin sister and I had been in several movies, as well as commercials, a music video, and had even done some fashion modeling. I still remember my mother's face when our cousins called her to say that they saw us on a huge billboard in New York City! You would think that with all that experience around the camera I would be very comfortable with public speaking. That couldn't be further from the truth. I feared being judged by others even more so than most people. The *social evaluative threat* component of public speaking fear, is perhaps the strongest single reason that you were motivated to pick up this book.

This fear of the spotlight isn't just in your head. In fact, there are a lot of scientists who have been studying the effects of social evaluative threat on the stress response. Back in the 1990s, a research lab in Germany developed a study procedure called the Trier Social Stress Task (TSST). It's considered the gold standard in stress research, and has been implemented in over 650 published studies.

Imagine for a second that you signed up to participate in one of the studies I just referred to. You are greeted by a researcher and escorted to a room where three judges sit in white lab coats and stare poker-faced at you. You're asked to sit down. There is a video camera and a tape recorder on the desk. You're asked to stand at a microphone and deliver a brief speech. Then you are told to pretend that you're a job candidate and you are supposed to convince these three judges that you are the perfect fit for this position. They tell you that you have ten minutes to prepare this speech, and ten minutes to give the speech. You're also told that the judges are specially trained to monitor nonverbal behavior, and that a video analysis will be conducted on the recording of your talk.

Are you feeling nervous yet? You may even be sweating just reading this, if you're still terrified of public speaking. If you are feeling that way, imagine sitting down to prepare that speech, knowing that in just ten minutes you're going to be recorded and judged in detail.

Once you are left alone to jot down notes and prepare your talk, you scramble to think about all the awesome things that would make you an excellent candidate for this pretend job. You notice that you're sweating and you don't even know why you're so nervous. It's not like these judges have any kind of effects on your career or your life. You manage to collect yourself a bit, and hold your head high while being escorted back to the judges in the other room.

They sit there in their white lab coats, just staring at you. To your surprise, they tell you to hand over your written notes and that you may not use them in your presentation. Everything stops! The panic really starts to set in.

You begin your speech, a bit shaky and groping for words. After what seems like a decent amount of time speaking, you are pretty much out of things to say. One of the judges says, "You still have some time left. Please continue." You try to think of anything you have left out, or any new point you could add, just to fill some time.

After some heroic rambling, you putter to a stop, certain you have at least exhausted the minimum time. But the committee just sits there staring at you for what feels like forever. Their silence is actually the standard 20 seconds of silence they use in each case when a participant thinks they are finished before the ten-minute mark. At this point they begin to ask you questions and you do your best to answer each one. Eventually, a timer goes off. Finally! You can now put this nonsense behind you.

Nope. Think again. Just when you're ready to run out of there, the judges ask you to begin with the number 1,022 and do serial subtractions of 13. Is this a joke? You start out….1,022, 1,009, 996,….You keep going but after a few of these you get interrupted by one of the judges who abruptly says "Stop. 1,022," indicating that you need to start from the beginning because you made a mistake.

What I just described to you is the standard procedure for the TSST, which has been modified in various ways to analyze many different aspects of the stress response, since its inception several decades ago. Most of the studies have included having participants spit into a tube to measure our two friends, noradrenaline and cortisol, through biomarkers in their saliva. I've learned a lot during my years of administering the test to my own research participants—both about stress in general, and about my own fear of public speaking in particular.

Now that you're familiar with the key players (noradrenaline and cortisol) in your fear of public speaking, and you know how different elements of public speaking influence these key players, we can start to get to work. I'm about to share with you what I learned so that you can apply these techniques when it comes to conquering your own fear of public speaking.

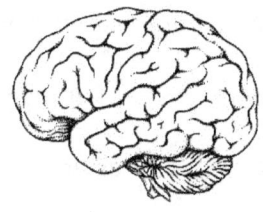

"Fear has two meanings: 'Forget Everything and Run' or 'Face Everything and Rise.'

The choice is yours."—Zig Zigler

## Chapter 3: Science-Based Tips to Combat Fear

If you were stressed out just reading the description of the TSST, you are not alone. In my first year of graduate school, there was a journal club that we were encouraged to attend. In these club meetings we had to read a neuroscience article and then discuss it with our peers. I remember my heart would start to race while we were just sitting around the table, when it was almost my turn to speak. At times I just wanted to get up and run out of the room.

By my second year of graduate school I had begun to focus my research on our friend noradrenaline. I was buried in scientific papers where researchers were testing the effects of different drugs on various brain regions in rats. In several studies, scientists had injected a drug directly into a portion of the rat's brain involved in fear, called the amygdala. When a rat's "fight or flight" system gets turned on, the amygdala is activated and the rat freezes, his pupils enlarge, and he avoids anything previously associated with the mildest amounts of stress or fear. I know—these studies weren't always nice to rats.

One of the reasons scientists like to use rats in these types of experiments is that rats tend to fear novel environments. If you put a rat in a cage or apparatus for the first time, his noradrenaline will sky-rocket. If you repeatedly expose the animal to the same environment, the rat's noradrenaline response will decrease over time, as the rat "habituates," becoming accustomed to the once-novel environment. By varying the intensity of the stressful situation rats are exposed to, scientists have been able to measure the amount of noradrenaline produced directly from the amygdala. There is a tight correlation between intensity of stress and the increase in noradrenaline in the amygdala.

You can give the rat a drug by injecting it directly into its amygdala or into its bloodstream elsewhere in the body, and either way it will block the physiological and cognitive effects of its "fight or flight" response. The most common drug used in most of these studies has been propranolol (not to be confused with propofol the drug that killed Michael Jackson). This kind of drug is called a beta-blocker because it blocks noradrenaline from binding to receptors. The effect is similar to preventing a key from entering a lock. Without being able to bind to receptors in the brain, noradrenaline cannot do its job. Thus beta-blockers, by blocking noradrenaline from binding to receptors, can prevent the unpleasant physiological responses that we have regarding public speaking, such as anxiety, fear, and panic.

By my third year of graduate school, despite having had many opportunities to speak, my discomfort with public speaking wasn't going away. I realized that if I didn't learn to conquer it soon, I would seriously need to consider a different career path, since for researchers, successfully presenting your research at conferences is an important part of academia. Naturally, our faculty encouraged us to give presentations as much as possible during our graduate career, in order to prepare us for this aspect of the job.

The thought of talking to my peers was still scary, but I went ahead and submitted a cool study I'd been working on to the annual Cognitive Neuroscience Society meeting. I was actually pretty excited by the possibility of presenting my own study, but also in the back of my mind, I pictured my peers judging me as I stood next to my poster, like I didn't deserve to be there.

A few months after I submitted my abstract online, I received an email from the conference administrator stating that I had been selected for a prestigious award, "Graduate Students Present." The award would cover my travel expenses to the conference, and asked that I give a live presentation in front of attendees—instead of just standing next to a poster on my research, which is what I had been expecting. I had planned to take a small step toward facing my fear of public speaking, but now it seemed as though I would be taking a much larger step in front of an audience of other neuroscientists.

Within seconds true panic set in. What was I going to do?

I suddenly remembered those experiments where rat researchers had used propranolol to block noradrenaline. Neuroscientists had done this in humans, as well. My mentor had even published a study on the effects in humans of propranolol and noradrenaline, in the prestigious journal, "Nature." I made an appointment with a psychiatrist. I went to see him and told him that I had a phobia of public speaking. I asked him if he would prescribe propranolol to me for giving talks because it was essential to the success of my career. Just like that he wrote me a prescription for propranolol. The problem of over-prescribed medications in this country is real, but I was glad to know that I would have propranolol to get me through the day of my talk. I thanked him and left.

Since I had studied propranolol, I knew a lot about it. I knew exactly when I needed to take the pill, so that the drug would be in full effect right when I needed it. Propranolol has a 1-4 hour half-life, meaning the pill would be entirely out of my system within 8 hours, but most likely much sooner. I took the pill 90 minutes before my talk just to make sure I was in the ball park of when it would be the most effective.

I also knew that the drug would be more effective if I consumed it with protein-rich foods, so I loaded up on tuna salad, almonds, and cottage cheese that day, even though it was difficult to eat with my anxious nerves. Then I took the pill.

It was like magic. I remember sitting in the audience, waiting for my turn. No dry mouth, no sweating, no heart racing. Just calm. I told myself this was going to be an amazing presentation. When they called my name I walked up on stage. As the assistant got my PowerPoint set up, I raised my head and took in the crowd. There were a lot of people, and, despite the propranolol, it was still scary as hell. For a moment, I feared the drug might not work and that old panic began to set in. So I took a deep breath and pictured how the drug was binding to the receptors in my brain. I knew, based on my knowledge of neuroanatomy, that the propranolol should be blocking my brain from taking up the noradrenaline, and thereby quelling the fear that I would normally be feeling at this moment.

As I spoke, the words seemed to flow effortlessly off my tongue. I couldn't believe it. I didn't stumble once. I held my head high and kept charging ahead. When my fifteen-minute presentation was over, my colleague came up to me and said, "That was amazing! What did you do to prepare for that talk?" Others came up to me and said I had delivered a great talk that day. It was magical. Everything about that afternoon was—except for one thing. The whole time I had I felt like I was watching someone else give the presentation—almost as if I were viewing myself in a movie. After the talk, when I should've felt elated, I didn't feel anything at all. I was completely apathetic about something that should have meant the world to me. And I was in a haze for several hours until the drug completely wore off.

My friends wanted to go out that night, but I just wanted to curl up and sleep in the hotel. No one understood what was wrong with me and why I seemed so emotionally flat. It was such a contrast from my natural bubbly personality, and I hated feeling this way for hours after my presentation. Was this the solution to my fear of public speaking? Not if I was going to speak more often than every couple years! The propranolol did allow me to give a great presentation that once—but going forward, I knew I would be fearing the propranolol experience, as much as I would be fearing the public speaking itself.

    If I was going to be speaking on even a semi-regular basis, it was clear to me that I needed more than just a band-aid approach to conquering my public-speaking fear. I was going to have to do a lot of work if I was ever going to truly slay that dragon!

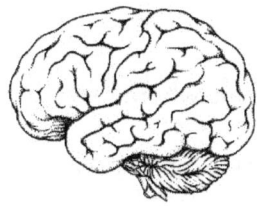

"You gain strength, courage, and confidence by every experience in which you really stop to look fear in the face."

—Eleanor Roosevelt

## Chapter 4: Science-Based Tips to Build Power

Where would I begin with my work? In my mind, I started going over all of those animal studies where researchers had habituated the rat to the novel environment, by exposing him over and over again, until his noradrenaline decreased. I thought to myself, what if I could do that to myself? What if I could somehow determine the scariest components about being in that spotlight, and try to "habituate" myself to them?

At my university, the lecture room in which great neuroscientists from across the country were invited to speak, was the same room where graduate students eventually defended their doctoral dissertations. Consequently, when it came time to see if your doctoral committee viewed your research as worthy of Ph.D., it was especially easy to feel intimidated presenting in this particular room.

Actually, I was terrified of that room. Since doctoral defense presentations are open to the general public, anyone and everyone can attend your talk, as well as ask questions after. Next, your doctoral committee of research advisors grill you to see if you really know your stuff. You then wait outside the room, while they come to a decision.

If all goes well, they announce you to the world as a Doctor. Knowing what was at stake in the preparation for my doctoral defense, and not wanting to dive into the propranolol robot sea, I decided to skip the meds and try to face this head-on, sans self-medication. I mean, you only become a doctor once; I didn't want to miss feeling that "high" by talking propranolol.

So first things first. I had to overcome my fear of that room.

Several days before my defense presentation, I asked the department administrator if I could use the keys to that room. The first day I just walked around the room to familiarize myself with it. I looked at where the light dimers were, and adjusted them. I looked at the computer and screen setup. I stood at the podium and practiced standing up tall and taking deep breaths. I walked the stage.

I also imagined what the room would look like when it was full. I looked out at the imaginary crowd; I filled them with as many people as I could think of. Some were mentors, some were colleagues, and some were smiling family and friends. I closed my eyes and tried to picture their faces. I didn't say a word of my presentation. I took a seat in the front row, and imagined being one of my audience members. I pictured myself up there, confident and speaking passionately about my research. I tried to imagine what I would be thinking if I was one of my family members or friends in the audience.

I started to relax. This didn't seem so bad. It actually felt kind of fun. I got up and sat toward the back of the room and imagined myself in the audience again. This time I tried to pretend I was an older faculty member with a lot of experience and knowledge of the field. Again, I started to relax, as I realized that they would probably be focused on the content of my research, which I knew was solid. I simply had to find my strength in the delivery. The more I visualized my audience and myself speaking, the more confident I felt. Feeling satisfied after just about fifteen minutes of my time, I returned the key and left. The administrator seemed surprised that I was in there for such a short amount of time.

The next day I did the same thing, only this time I stayed a bit longer and I began to walk around the room more like it was an extension of my home. My confidence grew as I started to get more and more comfortable with this room. It was no longer this scary place reserved for the elite speakers in my field. "This is my stage," I remember thinking to myself. "Tomorrow, this room will be my battlefield and I will slay my dragon."

The next day, on the morning of my dissertation, I delivered the best presentation of my life, medication-free. I captured the audience with my passion and I dazzled them with my confidence. I couldn't believe how powerful a couple simple techniques had been; just by getting comfortable in the room and visualizing my audience—i.e., "habituating" myself to my anticipated environment"—I had been able to calm my nerves for the day of the talk. Having gone through the habituation techniques, I literally felt like I was speaking in my living room, and these were just guests I had invited into the comfort of my home. The best part? I didn't need to take a pill to achieve this.

Being able to physically go to the room I would be speaking in, was a big help. Part of what was helpful about being there in the flesh was that I didn't have to imagine quite as much—the room and seats were all there in front of me, so in my imagination, I just had to fill the seats and pretend it was the day of my talk. It's true that whenever I can't physically visit a place that I'm scheduled to speak at beforehand, the visualization process is more difficult—but only in the sense that I have a lot more imagining to do. In those cases, I have to imagine the look and dimensions of the room, the lighting, the view from the stage, etc., in addition to everything else—but usually I've found that with just a little work, I can get quite a bit of information about the room, to help me keep my imagination tethered to the actual facts of the location.

Typically, I start by looking for pictures online. It's often the case that you can find quite a few images of a specific room, whether it be a hotel ballroom, a conference room, or a lecture hall. If your talk is part of a larger conference, you might have to do a little research to find out what specific room you are assigned to. You might even be able to find a video on YouTube that was recorded in the room you will be speaking in. Also, if you are unable to find satisfactory pictures or videos, you can always call and ask for the dimensions and maximum occupancy of the room, which can help you at least feel like your upcoming location is not a total mystery.

If you happen to arrive a couple days before your talk, ask the administrative staff if you can practice in the room after the sessions are over for the day. This isn't as uncommon as you think. You just need to ask. If you really feel weird about this, you can always check the schedule, attend the last talk for the day in the room you'll be speaking in, and then just linger after people leave.

I mentioned that my dissertation talk was the best presentation of my life at the time I gave it. Since then, I've learned quite a few additional science-based tips that have brought me to where I am today. But I've also learned a lot more about the human brain, and about the neurobiology behind just what makes public speaking such a source of anxiety for so many people. You may relate to this next one.

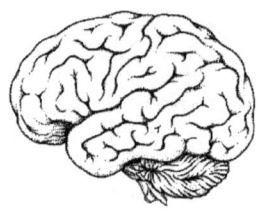

"Success comes from having dreams that are bigger than your fears."—
**Terry Litwiller**

## Chapter 5: Guilty Culprit #2—Drawing a Blank in Front of Everyone

Maybe you've had some positive experiences with public speaking. Perhaps you've even learned how to turn that nervous energy into enthusiasm. You get up on stage, grab the microphone and fire away. And then suddenly, in mid-sentence, you hit a silence that feels like forever... Who is the guilty culprit here? You guessed it...our fair-weather friend, cortisol. This guy can be a real pain in the butt. Remember how I explained that cortisol can actually enhance your memories? That depends on the timing of when it's released. We'll address cortisol's memory enhancing powers in chapter 9. When you're trying to recall important information in the middle of a talk, it is best for your cortisol to take a walk because he can block your ability to retrieve previously stored memories.

Hundreds of studies in rats and humans have demonstrated that cortisol blocks memory retrieval when levels are high. However, there's one key component. Remember our buddy noradrenaline? As you recall in Chapter 2, when I blocked noradrenaline with propranolol, this prevented cortisol's influence on my ability to retrieve stored information.

I had implemented the findings I had read about in countless animal and human studies where propranolol prevented cortisol from blocking memory retrieval. I can't even put into words how comforting it felt to be up there on stage, completely fearless of drawing a blank, because I understood the science behind what I had done to prevent this from happening. I felt like I was on top of the world and my confidence was so strong.

They say certainty builds confidence, so if you can increase your certainty about the delivery of your speech, as I did, your confidence is sure to follow and to grow exponentially. Except I'm not going to suggest that you take propranolol. Instead, I'm going to take you through some science-based exercises that will lower your noradrenaline. Remember that this is key. When it comes to cortisol, this is the stress hormone that is primarily influenced by social evaluative threat. That is one component about public speaking you won't be able to change. Most talks—whether they are live or online—are in front of an audience. Lower your noradrenaline with these tried, tested and scientifically-backed techniques, and you're good. First, we'll address ways to lower baseline noradrenaline, so you can begin your talk as relaxed as possible. Then we'll discuss ways to keep noradrenaline in check throughout your speech.

**Controlling Baseline Noradrenaline Levels**

<u>Tip #1: Habituate to Your Speaking Environment.</u> I mentioned earlier that I habituated myself to that big scary lecture hall where I had to give my doctoral defense. You can do the same by trying to learn as much information about the environment that you're going to speak in as possible. And believe me—with a little creativity you'd be surprised how much you can find out!

Remember that what is key in lowering your noradrenaline, is to visualize the room you will be in, and pretend you are there in front of a full audience. Base the size of your pretend audience on a reasonable guess, depending on the information you have. Then try to actually picture the *kinds* of people who will be attending your talk. The more details you can visualize the better. This leads us into Tip #2.

<u>Tip #2: Visualize Yourself Speaking from *Two* Perspectives.</u> I know that sounds weird, like I'm asking you to have an out-of-body experience or something, but I really want you to try this. It worked for me, and I'm confident that it will also be helpful for you. Whether you are able to do

this in the actual room you will be speaking in, or whether you need to just imagine it, go through *two* visualizations.

First, imagine that you are standing where you will be giving the presentation (a stage, platform, podium, etc.) and picture yourself talking from your own position, and looking into the faces of the audience. Get centered in your knowledge and allow your confidence to build. Remember, you are the expert or they would not have invited you to speak.

Once you've done this and you feel comfortable picturing your audience without the jitters, go through the visualization a second time. This time pretend you are one of your audience members. If you are in the actual place where you will give the talk, take a seat and try to imagine what your audience would be thinking. If you really want to get the most out of this exercise, select the most intimating person you can think of who will attend your talk (for me it was the department chair), and try to imagine what kinds of questions they would have. Allow yourself to become immersed in the *content* of your talk, and you will see very quickly that they will most likely not be judging you on your appearance, your speaking mannerisms, or whether they detect shakiness in your voice. Your audience is genuinely interested in what you have to say. So stand tall and say it.

## Controlling Noradrenaline Levels During Your Presentation

Those are great tips for lowering your noradrenaline levels before you begin to speak, but what about throughout the talk? We know that most likely—if you are human and have a heart-beat—the social evaluative threat component will induce a cortisol response. So as long as you keep noradrenaline in check you won't draw a blank. Let's talk about some tips throughout your talk to lower noradrenaline. The first one might make you laugh, and while I don't have the scientific studies to back this one, I encourage you to try it.

Tip#1: Hold a Mug in Your Hand While You're Speaking. In graduate school, I had a colleague who did something brilliant throughout her talks. I always thought she was the most confident speaker I knew. What she did is so simple, you might even laugh. She held a mug full of tea in one hand as she spoke. Whenever she started to feel nervous she picked up the mug. Have you ever seen someone holding a mug of coffee or tea when the person didn't appear completely relaxed? Picture it for a minute. When you're holding a mug it gives the appearance that you are calm, collected and relaxed. Tea is "sipped" for a reason, because it's considered a leisurely activity where people are relaxed in social settings. It's a powerful image to your audience, to do this in a professional setting, and as my colleague shared with me, it should make you feel more confident. Don't like coffee or tea? That's okay. Just fill it with water or whatever you prefer. The key here is that the mug conveys a message of relaxation more than a simple cup, or even a water bottle.

<u>Tip #2: Stop and Take a Sip.</u> Whenever you feel a bit of nervousness returning, and you need a few seconds to collect your thoughts, actually stop, pick up your mug and take a sip. The time it takes you to take a sip might feel like a long pause, but to your audience it will seem completely normal. If you really are starting to draw a blank I want you to do something that seems really scary. Instead of frantically searching for what it was you were about to say, and stressing out about it, allowing cortisol levels to creep up, when you take that sip allow yourself to daydream for a second or two. I'm not kidding. Just mentally drift for a few seconds. In those moments you will relax and whatever you were trying to remember will come back to you. Take that, Mr. Cortisol!

**Speaking of Cortisol—How Can You Reduce Your Cortisol?**

For a moment, imagine someone who is slumped over, arms folded and looking at the ground. What is this kind of body language communicating? It's not surprising that you'd probably think this person is not very confident and that they're insecure. Now picture someone in a strong pose that conveys confidence. What does that look like to you? Standing tall, head up, arms raised. Right? Think about athletes from around the globe. What is the common pose that they make as they cross that finish line? Their head is held high, and their arms are usually raised in victory, otherwise known as a power pose.

A recent study by Dr. Amy Cuddy demonstrated that holding a power pose for just 2 minutes significantly reduced levels of salivary cortisol and raised levels of testosterone (a power hormone) compared to those who held poses of body language that lacked confidence prior to a mock job interview. Perhaps more interesting, the mock interviewers were unaware of what condition the participant was in (power pose versus insecure pose), but they only wanted to hire those who had held power poses. So try it yourself. Before you begin your talk, go to the bathroom and just hold a power pose for 2 minutes. You can do this several times if you'd like. You'll walk out there with some presence, and feeling like you own the place. You'll be ready to take that stage!

**Controlling Cortisol Levels During Your Presentation**

What if no matter how hard you try, your thoughts start to focus on how you're being evaluated. If this happens for an extended period of time, there's a good chance Mr. Cortisol will attempt to steal your show. What can you do about this? Try elements of the power pose. I don't mean throw your arms up in the middle of the talk, unless there is some way to do this within the context of your presentation. You'll find that you will prevent cortisol levels from rising throughout your talk, simply by standing up straighter, holding your head a bit higher, and keeping your arms open (versus closed) over your chest.

**After You've Nailed a Talk Allow Your Noradrenaline and Cortisol Levels to Sky Rocket**

You've just delivered an exceptional performance. My friend, you left the audience inspired, moved, and motivated. So much so that several of them approached you after your talk to express how impressed they were. When this happens, allow yourself to get excited. Let yourself feel that rush of adrenaline, and take a few moments to really let it sink in. Focus on their facial expressions as they offer their praise to you, and let noradrenaline burn it into your memory.

Why? Because you want this powerful image to make its way to the forefront of your mind, any time you get nervous in the future when preparing for a talk. In fact, before each presentation, allow your mind to flood with these positive images of praise that your audience offered. This will help you build more confidence, and for me this opened the gateway to a successful path without the use of anti-anxiety medication. Let's see if you've experienced this next fear. It's more common than you think.

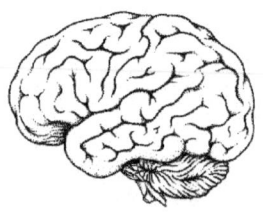

"Imposter Syndrome is the province of the successful, of the high-achievers, of the perfectionists."—Kate Hilton

**Chapter 6: Guilty Culprit #3—Imposter Syndrome**

Another fear that people sometimes struggle with is that your audience will discover that you're not qualified to speak on the topic that you are delivering. In those moments when you're fidgeting, or in your pauses, they'll suddenly be stricken with the realization that you are not an expert in your field. I mean, never mind the hours you poured into learning, the dates you blew off, and the cups of coffee you consumed just to burn this information into your brain. Never mind long nights in the lab or late evenings in the office comparing pie charts and excel spreadsheets full of data. You're a fraud and they know. They can see it now. You can no longer maintain this cover.

As wild as that sounds, most people can relate to what I just described on some level, whether you're at the top of a multi-level-marketing company, or the founder of your own online business. If you have suffered from those symptoms, you have experienced what psychologists call imposter syndrome. Those who suffer from imposter syndrome might feel as if their confidence is pretty average when they wake up in the morning, but by the end of their daily routine, it's a shriveled mess.

This is because if you have this syndrome, then all day long, your brain is second-guessing everything you say and do. It's as if you're watching yourself and evaluating every action from the perspective of your peers. So when you then have to give a speech, this can be your worst nightmare, mostly because the scrutiny you put yourself under 90% of the time gets magnified by about 1,000 in your mind the second you speak to your peers, students, or employees.

There are whole books on the topic of how to overcome imposter syndrome. For now I'm just going to share a few tips on how I learned to deal with this. One day in the middle of graduate school I remember having a lunch meeting with my mentor. He's pretty famous in my field and his work on gender differences and memory has received a lot of media attention. Yet, he too admitted that he had at one point suffered from imposter syndrome. He said it seemed to be a normal part of earning a Ph.D., especially in a scientific field, because you start to believe that you're supposed to be omniscient—that somehow you're supposed to know anything and everything about your field, down to the tiniest detail. In reality, there's not a single person on the planet who knows everything about a subject, even the ones who specialize in a certain area. Over the years, I've come to appreciate that this is part of the beauty of becoming an expert in an area. You've learned so much, but the more you learn the more there is to know, and as long as you stay excited and curious in your field, no one can fault you. The truth is, they actually don't expect you to know everything. *You* expect you to know everything.

I ask you to stop wherever you are in your career right now and assess what you know and what you have accomplished, by your standards alone. For a few moments, mentally take inventory of the last few years of your life and see how far you've come. Allow yourself to recognize just how much you actually know and get excited about what you don't know! As I started doing this more and more over my academic career, my confidence grew and so did the way I carried myself. Something else grew too. My voice. I was able to raise my hand and actually answer questions in class or share my ideas in meetings and journal clubs. The more confident you feel the more you will display this confidence in your body language. Others will begin to trust your expertise in an area because you've earned it and you've demonstrated it. Nothing has changed except your behavior and your confidence. When this happens, I want you to pay attention to the way others respond to you, and before you know it, this imposter syndrome will have to leave and go terrorize someone else, because it won't stand a chance with you.

One more important point: If you are invited to speak somewhere, you are being treated as an expert in the area you're speaking on or they would not have invited you. Minutes before I defended my dissertation, when I was trying to calm my nerves, my sister said something I would never forget. She said "You know more about your topic than anyone in that room." She was right. I was talking about my own research, and all the years of long hours of work came flooding back to me in that moment. It hit me then, and I've said that phrase to many others since then, right before they have spoken at an important event. It is so true. If you are invited to speak, you know more about that topic than anyone else in the room. It doesn't matter if there are 20 people or 500. You have something unique to offer that no one has heard of, or perhaps even thought of before. So own that! Every time before you go to speak I encourage you to say that mantra to yourself ("You know more about your topic than anyone in that room").

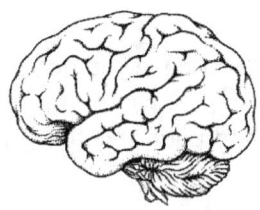

"There are always three speeches for every one you actually gave. The one you practiced, the one you gave, and the one you wish you gave." — Dale Carnegie

## Chapter 7: Guilty Culprit #4—Perfectionism at its Best

Even the most polished and experienced public speaker might struggle with this fear. You have memorized what you want to say. You have practiced it over and over and there are certain elements that you really want to communicate in a particular way. You begin to be so focused on *how* you are delivering your message that you lose sight of the message itself.

By no means am I suggesting that you shouldn't rehearse what you want to say. The more you practice the more confident you will feel. But how do you get over trying to say something perfectly? Do you ever put off a presentation because you think that you won't say it *exactly* how you practiced it? Here's my advice. If you've got certain elements that are essential to the style or the content (for example, a joke that you came up with that fits well with the material, or a vivid analogy that really emphasizes a key point) then take pen to paper and jot those down. There are scientific studies that support the notion that writing something down will help you remember it. Keep those points on a flash card and carry that card with you. From time to time look at the card again.

Each time you look at it you will be reinforcing those key points in such a way that they will subconsciously be there when it comes time to deliver your presentation. Aside from those points, as difficult as this is, I want you to let go. Just relax and allow yourself to say things a different way each time you rehearse, no matter how much a specific phrase sounded cool, or how much you felt like that is the only, or the best, way to convey your information. When you do this, it will become impossible to hold onto any specific way of saying something because it will be different each time, but the core of your message will be the same.

Focus on the content, and reconnect with the passion that you feel about your topic, as if you can't wait to tell a friend. Allow yourself to lose yourself in the content. When you do this, your delivery will come out in a more polished and articulate way without even trying, because you will naturally be more connected to the content you are communicating. Try this. Every time you rehearse and get caught up in the way you said something, I encourage you to push that out of your mind and return to the core of your message. Know that your choice of words may be different each time you rehearse and between what you rehearsed and what you actually deliver, and give yourself permission to do this. Remain focused on your message.

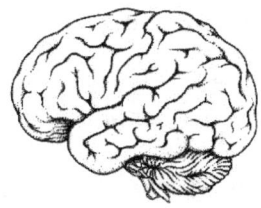

**"Give more to your audience than they have any right to expect."—Tony Robbins**

## Chapter 8: Overcoming the Panic Recap

By now you've learned a handful of science-based tips that you can use to really make a difference any time that panic starts to creep in. It will take some time and dedication, but if you're willing to put in the work, you should begin to notice a difference right away. The more you apply these techniques, the easier it will become to apply them the next time as well.

When I was applying for my postdoctoral fellowship I had to give a job talk. I applied these techniques and I was able to convert my nervous energy into enthusiasm. I was thoroughly prepared in every way…except for the fact that I had not estimated enough time to make it to the interview given current traffic. After frantically cutting off several people in my haste, I pulled into the parking lot at 8:58—just 2 minutes before I was scheduled to begin!

My heart was racing and I started to sweat. Not a good look for a job interview. I'd never actually been late to an interview before. I couldn't believe this was happening, but there was nothing I could do to change it. To make things worse, I couldn't find a parking spot. It was 9:03

when I finally found one. I raced into the elevator and when I got to the 3rd floor it was 9:08. When I finally made it to the suite it was 9:10.

So much for controlling my baseline noradrenaline levels. I tried to collect myself, but I was pretty sure they wouldn't hire me. They did not look pleased when they met me, but they ushered me into the conference room and helped me set up my PowerPoint. In those moments I could've crumbled and completely fallen apart. I could've let my insecurities about the way they would perceive me affect my presentation, but I ignored those thoughts. Instead, I got excited about my presentation. I connected with the content and I sincerely wanted to bring my message to them, whether they were planning to hire me or not.

Everyone will most likely experience pitfalls like mine that could derail you. In those moments you have to commit to why you are there and what your message is about. I ended up finding that spark inside me, and gave a presentation that left my audience speechless. They told me afterwards that they were not pleased with my being late to the interview, but that my presentation spoke much louder about the kind of researcher I am and the quality of my work. With that praise, they hired me.

By now I'm pretty sure that if you merely wanted to avoid feeling panic when it comes to public speaking—as opposed to really becoming a good speaker—you would not have picked up this book. You want to be a powerful speaker with a presence. You want to captivate your audience and leave them speechless, as I did in that job interview. Now that we've laid the groundwork for kicking panic to the curb, let's apply the science we've learned to building your presence and your power.

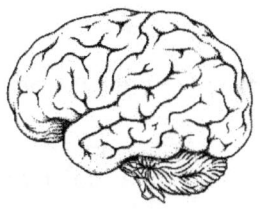

"Speech is power: speech is to persuade, to compel, to convey."—Ralph Waldo Emerson

**Chapter 9: Let Your Power Build**

Giving a presentation that was so strong that it landed me a job even though I was late to the interview, definitely helped my confidence with regard to my abilities as a public speaker. In subsequent talks, I presented with a bit more strut in my step—convinced that I had gotten the hang of public speaking. When it came time to give my next job talk, there was much more on the line. I had been invited to interview for a research professorship at Arizona State University. I was asked to prepare a 60-minute talk and I spent hours selecting just the right slides and images to convey my message. I flew from California to Arizona and I remember feeling a hint of nervousness since this was an exciting opportunity in my academic career. Right before the talk I excused myself to the restroom and in one of the stalls I practiced my power poses.

Instantly, I relaxed. Head up and shoulders back, I walked back into the conference room to find a whole lab of people I didn't know surrounding a long wooden table. I quickly set up my slides, while rehearsing my presentation in the back of my mind. My excitement grew as I did, and before I knew it, I felt as if I was here to share some great news with a friend. This would be a breeze.

Then, in the middle of my presentation, the projector suddenly malfunctioned. Just like that, my hours of hard work were in jeopardy. But I was so caught up in the excitement of what I was sharing, that I almost didn't notice. In the face of this technical difficulty, I rose to the challenge and continued the presentation without my slides. I told my audience that I would email them the slides after the talk, and then edited my content on the fly as I went through it, only describing what they would have seen on the slides, in the handful of cases that the material absolutely necessitated it. The talk went great, and the next 45 minutes were over in the blink of an eye.

After my talk, several members of the audience came up to me to tell me that my presentation was one of the best they'd ever seen. In those moments I allowed my noradrenaline to soar. I was on top of the world. I got the position.

A big part of my success in this situation was due to my understanding of how to make a stressful situation work for me—or, to say the same thing in terms of neurobiology: how utilize noradrenaline and cortisol to work in my favor. Let's dive in.

**Power Tips in Preparation**

We've all heard the phrase "fake it till you make it." In my experience, this is an effective and somewhat easy approach to building confidence with public speaking. I know what you're thinking. Didn't she say earlier how important it is to be authentic and honest during the presentation? Yes, that's absolutely true. That's why this tip comes with a caveat. Do this in preparation for your speech, especially when you are just starting out with public speaking. Then, when you're in front of your audience, try to be as authentic as you can throughout your presentation. You may occasionally have moments of insecurity as all polished speakers do, and in those moments it may be helpful to incorporate parts of these pre-presentation tips, where they are appropriate.

Tip #1: Music. Find that song—you know the one. That one song that gets you pumped!! The kind of song that makes you feel like you can do anything. I don't need to tell you that there is scientific evidence that supports the power of music to create emotions, or place you in a confident state where you feel like you can conquer the world. But in case you're wondering, there actually is.

Neuroscientists believe that there are three main reasons for the universal emotional appeal that music has: 1) There is an emotional context that is associated with particular types of music. You may play a song that is subconsciously associated with joy, excitement, or fear, without even realizing it. 2) There are often memories about our personal experiences that are connected to specific songs or kinds of music. You may remember exactly where you were the first time you heard that song and what you were doing. Every time I hear Whitney Houston's "One Moment In Time," I remember crossing the finish line at the LA Marathon years ago, because I had it playing at the time. Songs that are associated with the Olympics are typically used to convey inspiration, pride, and success. Music associated with sports and winning can be a great confidence booster right before you are about to present. 3) Finally, music induces visual imagery that can be very emotional.

When we listen to music that is emotionally moving, images related to our own personal experiences flood our brains, and can often induce a very pleasurable feeling if the emotions are positive. In brain imaging studies, music has been shown to activate a key brain structure involved in reward behavior and drug addiction (the nucleus accumbens). My advice to you is to take a few moments to reflect back on your life and select at least three songs, and make sure that they each have *all* of those elements. Select songs that you associate with joy, elation, and motivation. Select a few songs that give you energy and remind you of a time when you achieved success or felt proud. Finally, select songs that are so powerful that they flood your mind with images that are inspiring and motivating.

Some people actually put together a power playlist and they don't listen to those songs at any other time. They bring out that list when they need to feel those positive emotions to get psyched up to go on stage: confidence, power, strength, inspiration. Try it. Either play one song over and over on a loop, or put together your own playlist and only play it right before your presentations.

<u>Tip #2: Your Bad-A$$ Role Model: Your Avatar.</u> Identify that one person (they could be famous or just famous to you), who you look up to and who embraces the kinds of qualities of a confident person. This is what we'll call your avatar. Picture them doing what they do best in your mind. If you can combine this with tip #1 that's even better! For me, when I think about the type of confidence I want to portray in my presentations, it's not some heroic scientist or academic.

My avatar is actually Beyoncé. To me, she embodies everything I imagine when I think of confidence, strength, and power. From the way she carries herself with her shoulders back and her head high, to the fierce look in her eyes when she performs. She has the sort of expression on her face that says "Don't mess with me" and "I am proud of who I am." This attitude is exactly what I want to embrace every time I step on stage to speak, or go on a live broadcast to address my market, or record a podcast for my audience.

Visualize your avatar and try to embody their attitude before you step onto the stage. Pretend that you are actually this person for a few moments. This also helps you take the spotlight off of yourself for just a bit, and it can be a lot of fun. I mean, how many times have you seen Beyoncé deliver a talk on noradrenaline and memory enhancement? Haha! Good times.

To get yourself in the mood for your presentation, I recommend combining Tip #1 and Tip #2. Whether you picked a musician or not, choose a song from Tip #1 and think of your avatar. Naturally, I selected one of my favorite Beyoncé songs, "Run the World." It's a female-empowering, upbeat, fun song to dance to. I would blast this song on repeat and just dance to it over and over to boost my confidence when getting ready to present.

You can't simply picture your avatar. You have to actually try to pretend that you are them. Allow yourself to embody that person in your mind each time you rehearse until you build your confidence. This is what I mean by "fake it till you make it." When you are practicing your presentation, once you get into your content, try to let go of the avatar and focus on your message. The trick is to not allow your brain to wonder how others are perceiving you. This is what can throw off even the most polished speakers. Whenever you start to drift toward thinking about how your audience is thinking about you, that's when it's time to go into "Beyoncé," or avatar mode, just for a few moments.

<u>Tip #3: The Power of Movement.</u> I've been drawn to self-development for most of my adult life. When I started my Research Professorship, I had the pleasure of attending an amazing seminar by one of my mentors in fitness and business, Chalene Johnson. Her seminar, "Smart Success," is one I highly recommend for business entrepreneurs. You can find my testimonial on YouTube. At this event I heard someone speak for the first time. When I look back on my path of professional/personal development, there was my life before I saw him speak…and my life after. Never before had I ever been so moved. As I stated in a text to a friend afterwards, "He shook my soul."

His name is Bo Eason. From the second he opened his mouth on stage I was mesmerized. I couldn't take my eyes off of him. It was as if my world stopped spinning. I know now, what it was that captivated me, but I didn't know it then. Bo Eason actually specializes in the power of public speaking, and sharing your personal story with the world in your business or professional endeavors. If you're interested in learning more about his incredible story and how you can have a more powerful stage presence, I encourage you to go to: www.boeason.com.

Among the most salient tips that he taught us that day, there was *one* that stood out above all the rest, and that is the power of movement. He explained that he had actually worked with a movement coach, who had studied in depth how predatory animals move.

I want you to take a few moments now, to imagine that you are a predatory animal. For the sake of simplicity, lets say you are a lion. When you go to greet your audience, continue to imagine that you are a lion and instead of feeling nervous about needing their approval, act like you are a predator on the hunt. It's a hunt to move toward them with the passion and conviction that lies at the core of your message. You are not a passive speaker. You are a crusader. Take a few moments to feel that.

If you think of someone who has captured the stance of a predatory animal what does that look like to you? Picture them moving on stage, maybe even going down into the audience. Where do your eyes go? Do they turn to your cell phone? No. Do they turn to the person next to you? No. Your eyes follow every step this person takes because they have a presence that makes it impossible to look away. That's how I felt that day, watching Bo Eason speak. An hour had flown by and it felt like I hadn't even blinked. When I got back from that seminar, I tried implementing this type of movement, confidence, and presence in my own body language everywhere I went. Just a few days after this seminar, I had been asked to give a guest lecture on my research on exercise and the brain to students in a psychology course at ASU.

I decided to make this lecture my best yet. I knew the material well, because I was lecturing on my own research, but I wanted to captivate these young minds. I wanted the material I presented to inspire them to go into research themselves. About 30 minutes before I left for campus I turned on my favorite Beyoncé song and danced in my living room. Then I closed my eyes and imagined that I was a lion. I know it sounds silly, but it really works. I suddenly felt taller, stronger, more alive.

I started walking around the room, and as I practiced my speech I took larger and larger steps. I stood taller with my shoulders back.

I never practiced a talk by sitting or even just standing again. I remained in constant controlled motion. It was in some strange way exhilarating. Have you ever seen someone when they are panicked and not sure what to do to solve a problem? They start pacing. This isn't exactly what you are doing, except in this case you are taking long, orchestrated steps and remaining in control. With each step you are focusing on what you plan to say.

Recent research suggests that when we move, we engage more of our prefrontal cortex. That's an area of the brain that is involved in abstract thinking, creativity, and planning. This is probably why, when people feel helpless or unsure of how to solve a problem, they immediately begin moving. We want to capitalize on this idea, and allow the motion to activate areas of your brain to get you prepared for a confident speech. I did this in my living room until I felt like I was on a high I couldn't come down from. I was so excited. I jumped in my car and drove to campus.

As I entered the room I saw young faces staring at their phones, some staring at their books and most of them looking bored. Instead of feeling afraid or intimidated by the daunting task of sparking their curiosity, I was up for the challenge and felt really excited about it. I walked up to the podium and set up my slide show.

I didn't return to the podium for the remainder of my lecture. I spent the entire time in motion. Within the first few seconds there wasn't a bored face in the room. I had their full attention. They followed my every move. Subconsciously, when you command the audience this way, it can build your confidence tremendously. I left the classroom feeling like I had definitely made a difference.

I had barely returned to my office before there was a knock at my door. When I opened it I immediately recognized the face of one of the most excited students I had just lectured to. He had sat in the front row and seemed completely engrossed in my lecture. I asked the young man what I could do for him. He said "Dr. Segal, I loved your lecture! I just found your research fascinating and I would do anything to work in your lab! Do you have any open positions for volunteer work?". Long story short, he ended up being my research assistant for two years, and one of the best I've ever had. As you build your power with public speaking you'll start to see some amazing opportunities open up for you.

**Power Tips During the Presentation**

Tip#1: Find your Anchor. There's usually at least one person in the room who will be completely engaged in what you're saying. They'll laugh at your jokes, they'll nod when you say things that resonate with them, and they usually won't take their eyes off of you. Whenever I begin a presentation I quickly scan the room to find this person. Then, whenever I have a moment of insecurity, I look back at them. This person is what I will refer to as your anchor. You want to keep scanning the room and connecting with your audience as you move and as you speak, but if you ever start to feel uncomfortable for a few moments, make eye contact and connect with your anchor. This will keep your noradrenaline levels in check, and prevent Mr. Cortisol from sneaking up on you. Sometimes you can be lucky enough to find several of these people. In that case, bounce around to each of them with your eyes, because most likely each one of them will offer some unique form of positive energy— whether it is smiling, or leaning forward with great interest. Sometimes focusing on all these positive aspects from your audience at once can really help build your confidence.

Tip#2: Move! I talked about the importance of movement when you are rehearsing, but the most important time to move is when you're up there speaking to your audience. Remember to picture yourself as a predatory animal from the moment you step onto that stage or hit "go live" on your cell phone. Remember that you are on a hunt! You have a very important message to bring to your audience and they can't wait to hear it.

If you are at a podium and you have notes there, step away from it as much as possible and walk around. Return to it only when you need to glance at your notes. If you are on a stage, make sure to walk from one side to the other throughout the presentation, and if you are on a floor that's level with your audience you may even want to walk into the crowd, up and down the rows. This actually closes the formal gap between you and your audience, so that you seem more approachable and authentic. You may want to work up to this one. I had a professor at UCSB who used to walk up and down the rows of our large 300-seat lecture hall. I remember feeling like he really cared about each and every one of us, simply because he did that. I was captivated from the second I sat down, and I didn't blink until the bell rung. It was such a personal approach that really had an impact on my motivation for that course, and I did well as a result of my engagement!

Along with movement comes the importance of those power poses that we discussed before. Make sure to try to incorporate them while you're on stage as much as possible. That doesn't mean sporadically jump into a victory pose while you're in the middle of transitioning between slides. It means to constantly be aware of your own body language and how you are feeling. Check in with yourself. Are you slumping over at all? Pull your shoulders back, hold your head high, and try to make sure that your arms are not folded, even if this is your tendency. This is a big one. As much as possible, avoid crossing your arms or doing anything with them that conveys that you are uncomfortable with who you are. You may think this seems silly, but experts who specialize in body language analysis suggest that folding the arms subconsciously tells your audience that you are protecting yourself or hiding something from them. Instead, try to use your arms to assist you with expressions involving the content matter and to emphasize key points.

Tip #3: Authenticity. Be You. Fitness has always been an important part of my life. During graduate school I was asked to be a runway model for a fitness-apparel line at an annual event in Orange County. While I was extremely excited, I was also incredibly nervous. I had done some modeling and acting with my twin sister when I was little, but this was different. I was alone, I was half-naked and I would have to strut my stuff with mega-confidence. Luckily, I was in excellent shape, but as even the best figure competitors will tell you, it is as if the audience is examining you and that your flaws are just out there for everyone to analyze.

I remember how much fun it was laughing and joking with the girls backstage while we had our hair and makeup done by experts. Then it was show time. We were all lined up backstage and my mentor, Chalene Johnson who was holding the event, walked up to us and asked if any of us were nervous. Each one of us raised our hands. I remember being a bit surprised by my nervousness. Why would I lack the confidence to just walk out there and dance to some music? For the same reasons that speaking in public terrifies people. It's because you feel like people are judging you and critiquing everything about you, starting with the way you look. This comes into play whether you are in front of a live audience, or live streaming on social media.

I'll never forget that moment when Chalene walked up to me before the fashion show. She put her hands on my shoulders, looked into my eyes with the fiercest sincerity and said, "These people came here to see YOU. Just be you, and have fun." Then she said to all of us, "Do you hear that crowd out there? They came to see you! Go out there and give it your best!" And with that I threw back my shoulders, raised my head, smiled and stepped onto the runway. I lost myself in the moment and it was one of the best evenings I've ever had.

Chalene's message stuck with me after that night and I've remembered it every time I've stepped on stage to speak. It wasn't even about being beautiful that evening. It was about bringing my energy to the crowd. That is what you need to remember each time you speak. No matter what your topic is, if you can bring your passion and energy to the subject you communicate, people will respond to you. They will be moved. The more you allow your true personality to show, and the more fun you have delivering your content, the more you are bound to move people.

Tip #4: How to Deal with That Difficult Question Asker. Congratulations. You've started to implement some of these tips to feel more confident and prepared prior to your presentation, and you know what to do to remain powerful throughout your presentation. But what about handling challenging questions? For live streamers, answering questions on the spot is essential to engaging with your audience and receiving more interaction from them.

I know that many people fear not knowing the answers to difficult questions, but it's more important to be honest and admit what you may not know, as long as you express concern about following up with answers. If you are a teacher, you know how much this matters to your students. They don't want to be taught the wrong information. When I was teaching a neurobiology lab during grad school, we were also part of committees that supported the other grad students' classes. We would attend and help with setting up and breaking down the lab, and also help answer any questions that the students might have.

One evening, I was teaching my students about a neurological surgery we were performing on rats, so that we could study the consequences of this surgery on the animals' behavior in future labs. The students had a question about the current state of research in an area, and I told them that I would look into the literature as soon as that session was over and email the class about what I found.

Luckily, a committee member volunteered to perform a quick literature search for me, while I was delivering my lecture to my students. He came back to my class in about 30 minutes with the information and the students were so impressed. They told me later how much it meant to them that I was honest, and that I didn't try to give them the wrong information that wasn't current. Kudos to my committee member for helping me out on delivering the current findings! I encourage you to be open about what you do not know with your audience in the same way that I was with my students. They will respect you and trust you even more for it.

Occasionally, you may be faced with someone who has ulterior motives for asking you a difficult question, and they're not actually interested in the answer at all. When you are faced with someone like this, they are trying to take the spotlight off of you because they want attention. Give them what they want, but do not give up your power. When I get a difficult question like that, and it is clear to me that the person is trying to put me on the spot, I acknowledge them by saying "that was an excellent question." Then if I don't know the answer I'll say "My years of experience tells me x, y, and z, but I'm not sure if that completely answers your question. Let me look into it and I'll get back to you." Or you can turn their attention to other resources that might have the information that they are looking for. The most important point is to make them feel important for asking their question (that's what they want), and then to make sure that you re-establish your power by emphasizing your experience, knowledge, background, or expertise.

**Power Tips for After the Presentation**

Perhaps the most important time to build your power with public speaking is actually *after* those presentations. I always follow the three Cs: Celebrate, Connect, Create.

Celebrate: Remember how we discussed the importance of focusing on all the positive feedback you receive after you kicked butt on a presentation? I'm literally referring to the minutes to hours following that presentation. This is because our friend (noradrenaline) and our frenemy (cortisol) will be like two toddlers jumping up and down throughout your talk, just waiting for you to tell them they can play.

You've been keeping them at bay the whole time, remember? Well as soon as your talk is over it's time to let them party. Your body and brain still recognize the importance of this event, even if you've been applying the tips I've discussed in this book. Your amygdala and your adrenals will naturally want to release these stress hormones. Allow that to happen *after* the presentation because this is when you want those subtle comments about how spectacular you were to be etched into your memory. Let your amygdala do its thing! Whether you realize it or not, over time these memories will build more and more confidence for the next time you give a presentation.

In addition to the neurochemical perspective, there is a large psychological component to whether you feel good about presenting again, so when you're reflecting on how you did, allow yourself to recognize your accomplishments. There will always be room for improvement, but often times the most driven and successful people do not take time to recognize and appreciate their journey. In the days that follow your presentation, treat yourself to something that makes you happy and allow yourself to accept that gift as a celebration. If you are new to public speaking, you may want to celebrate that first step, conquering your fear of getting out there to speak in the first place. Or maybe you are going live for the first time on social media. That is a big deal, especially when you are first starting out. For experienced speakers this is still important to celebrate, in order to remain enthusiastic and motivated with further talks.

<u>Connect:</u> No matter what field you are in, if you approach public speaking with the kind of energy, passion, and authenticity we've been talking about, then there's a high chance that people will approach you about opportunities after your presentation. They may or may not be in your

field, but that doesn't matter. I encourage you to connect with as many of these people as possible. If someone approaches you after a talk, obtain their contact information and follow up.

After that lecture at ASU, I had inspired that student so much, that he approached me in my office just minutes after my lecture. Other people may not be as courageous. They might email you a few days after your presentation, so make sure to be on the look out for opportunities like these. After I gave a research talk to healthcare personnel at Maricopa Hospital, a graduate student approached me and asked if I would like to be a Co-Investigator on a Mayo-Clinic funded research project that focused on nutrition, education, and exercise. I said yes!

Create: Anyone who has delivered an exceptional presentation feels an intense "high" right afterwards that probably lingers for the next few days. Capitalize on this "high" you feel and use it to drive you in your next business endeavor or important goal. This is where you can gain a lot of momentum with motivation. Even the most productive people have set backs, and we can't be on top of our game every second of every day. Given that you most likely presented on something you are passionate about, it's time to use this as fuel and nurture those relationships with people who have connected with you because of your talk.

Take action. This is when you do not want to wait. If you do, chances are the excitement and energy that moved people when they saw or heard you speak will die down over the next few weeks, mostly because people get mired in their own responsibilities and so you become put on the "back burner," for when they "get around to it." At the risk of sounding cliché, the best time to "seize the day" is right after you've presented. So connect with those you've inspired, capitalize on that connection, and create something—a project or a goal that builds on that presentation you delivered.

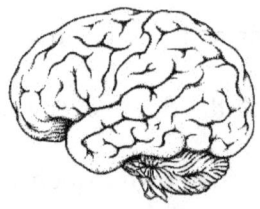

"All speaking is public speaking—whether it's to one person or a thousand."—Roger Love

**Chapter 10: Finding Your Power in Going Live**

It was 4am when my alarm rang that morning. I woke up after a deep and relaxed sleep, which given what I was about to do, would have never been possible several years ago. I would've laid there, wide awake for hours, staring at the ceiling. My noradrenaline and cortisol would've been like obnoxious roommates, partying and blasting music until the wee hours of the morning, keeping me awake. This is what happened in those earlier years of my career when I knew I had to give a talk the next day. Not a wink of sleep. This morning there was a lot more at stake than usual. I was going live on the Channel Five News at 5am to discuss scientific research related to a local sporting event.

As I was getting ready, I glanced out the window. It was still dark outside. As strange as it sounds, I felt like a child on Christmas morning. I felt giddy and excited, instead of being nervous. Many of the tips I've discussed in this book got me to this place. When I arrived at the baseball stadium I met the news team and we chatted and laughed. They went through the prompts and cues with me. At 3 seconds to 5am I heard them say, "And we're live in 3, 2, 1…"

I looked first at the camera and smiled as the reporter introduced me and then I focused on her as she asked me questions. Everything felt so natural and comfortable. I wasn't thinking about who could be watching this right now. I wasn't caught up in making sure everything sounded perfect. More than anything I was focused on my passion for this research, and the message that I wanted to bring to people to help improve their lives. Before I knew it we were done. What a rush that was!

Afterwards, the reporter said she really enjoyed interviewing me because I was approachable, and my passion for my research shined through. While I walked to my car I noticed that the first few rays of sunlight were starting to peak above the horizon. I realized in that moment how far I had come, from that terrified young woman embarking on a career path that would be laced with public speaking events, to someone who was confident enough to share her message live on the news.

Years later, my message is more meaningful to me and more powerful that it has ever been. I want more people to learn about the brain. If you understand more about how the brain works, you can be more efficient in changing behaviors and thought patterns that can help you achieve a desired result in your life. Today, I convey my message through streaming live on Facebook and teaching people about the brain in entertaining and interesting ways.

You may feel like you're at the point of your career where you have mastered speaking in front of live audiences, but speaking to a camera without seeing the faces of your audience is a whole new ball game.

Tip #1: Get comfortable. You know that feeling you get when you read or see something really fascinating and you're so excited that you can't wait to go tell somebody? That is the absolute best time to go live on Facebook, or any other live platform, especially if it is your first time and you're nervous about it. Some people announce when they are planning to go live to their followers, but I would not recommend doing this, at least in the beginning. This allows you to feel a bit more in control, to lower the pressure and make sure you're in the right mood to go live when you are ready.

When you go live on Facebook, the camera flips so that you actually see yourself as you talk, and there is currently no way around this. I struggled a bit with this at first, because it feels a little awkward and it makes it easy to criticize how you look. If you struggle with perfectionism you'll need to do your best to push past this and fight the urge to overanalyze yourself. You can practice in front of a mirror several times until you become aware of most of the things that might distract you as you're speaking live. You may make facial expressions that you didn't realize (e.g. raising one eyebrow). Take in every detail about your appearance as you practice in front of the mirror, because just being aware of these factors will help you feel more comfortable, and less distracted. One of the most common reasons that people give for not wanting to go live yet on Facebook is that they don't like the way they look on camera. If after practicing in front of the mirror, you still don't feel comfortable going live on your professional page, then try it on your personal page first. Allow yourself to engage with other people and continue to share your message or create funny/entertaining videos, to attract more followers to you personally. Then when you feel ready, jump onto your professional page.

    I would encourage you to always watch your videos right after posting them. You can learn a lot from observing yourself, and if you catch something that you really would prefer not to have in cyberspace you can delete it. If it's on your personal page and you're just starting out, I think it's fine to remove a couple of live videos until you feel comfortable.

However, I would suggest never deleting a live video from your professional page. It is often after the live feed that people will comment on the video and engage, because they may have missed the opportunity to view it when you were live. They may also share it with people, which will expand your reach.

I posted a video several months ago where the content was great, and my outfit, the lighting, and my message was on point. This one hair kept flying up and sticking straight out of the top of my head the entire time. I simply laughed and posted a comment making fun of it below the video. Being able to recognize what may have gone wrong on live video and then turn it into something funny or a learning experience for others is far more valuable than simply deleting it. And people will appreciate your authenticity.

Tip# 2: The Power of Authenticity. I once saw a sign that said, "May your life be as awesome as it appears on Facebook." It made me laugh, because Facebook is full of manicured pages and polished photographs. It seems like everything is "too perfect." While I love the emphasis on all the positive and wonderful activities people are engaging in, let's be real. Life isn't always that perfect. We don't share the photos of the coffee we spilled on ourselves running out the door, or photos of our children with food all over their faces. In fact, I'm not even sure we show our real faces anymore.

On the one hand, going live requires even more preparation than videos that you'll be able to edit later. You'll need to make sure that as much as possible you are in the best setting, with good lighting, clear sound, and no interruptions. Yet, there are many elements of broadcasting live that you won't be able to control and your audience knows this. To some extent, this is what compels them to watch and why live video is so popular on Facebook today.

One of the things that makes live video such a unique communication platform is that there are often factors you can't control. Embrace this and have fun with it. I once went live and wrote notes down on a board for people to see, and when I filmed it the camera was flipped so that all the words were backwards. Rather than delete these videos, just laugh at them, turn them into a lesson learned and maybe even add some humor or entertainment for your audience.

When you are going live there is a lot more at stake because you can't pretend to be someone else. Your real personality shines through because you can't edit what you are saying. In a sense, you are demonstrating authenticity when you go live and consistently deliver the same type of valuable content that you would present in a more polished format. What I mean by this is that no matter how many things may go wrong that are beyond your control, if the content is quality and reflects the value of the material in the videos that you have had time to edit, this strengthens the audience's trust in your expertise and your delivery.

It is important to keep in mind the purpose of live video. Think of these videos as trailers to a movie. You want to use Facebook live streaming to invite people who do not know you, to begin to follow you. It is best not to save your most valuable content for live streaming. You want to try to keep your videos brief (usually under five minutes) so think of using this platform as a way to increase engagement and perhaps send your audience to other platforms where you've invested the time to deliver more in-depth content.

When you become more comfortable with live streaming, be consistent and stream often. Try to broadcast live on the same days of the week if possible, and at roughly the same times. Once you are more experienced with live streaming, you may also want to send an email out to your followers ahead of time with the topic and the day/time you plan to go live, so that you can increase your engagement.

Do not let the fear of not appearing "perfect" deter you from going live. In fact, most people are more interested in seeing "real" people. For example, if you're a fitness professional you may want to go live after a workout when you're all sweaty, red-faced, and your make up is gone. Your followers will appreciate this much more than the airbrushed modeling photos that you posted. You may be worried that if you let your guard down and show yourself in a less than perfect light you will lose credibility. As long as you are well-prepared with the content that you deliver, the imperfections of going live should not interfere, but instead make you more easy to relate to.

Another key aspect about this platform is the fact that it allows the essential elements that distinguish you from others, to shine through all the pitfalls—from the hair that might be out of place, to the lighting that might not be perfect, to the dog running in the background. None of that matters if the content you deliver is of comparable quality to that of your more polished work. If you are able to do that, it will be like a singer whose voice is just as beautiful live, as it is post-production, in a studio. That's powerful.

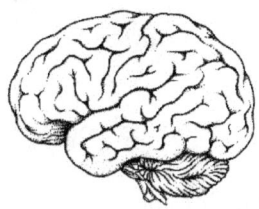

"The most precious things in speech are the pauses."—Ralph Waldo Emerson

## Chapter 11: The Power of Your Voice—Podcasts, Webinars, and Online Courses

With the recent explosion of podcasts for businesses, and the growing use of webinars and online courses for school, it is becoming more and more important to find the power of your voice. What do I mean by that? When you are speaking in front of a live audience, you can use your body language, and all of the other elements that we've discussed, aside from your voice, to convey confidence —especially if you incorporate the power poses mentioned previously. When you broadcast on a podcast you will not have these visual cues to accompany your voice.

Similarly, when we have no visual cues, gestures, smiles, or nods to provide interaction and feedback from our audience, presenting with platforms like webinars and podcasts can be more intimidating than presenting in front of a live audience. It's that much easier for your dragon to decide to make a guest appearance and make you think that everyone is rolling their eyes or laughing at you.

When the sound of your voice is the only instrument you can use to communicate your message, you need to make sure that you are feeling the power that you feel when you are up there on stage or in front of your audience, applying all the tips you've learned so far. About a year ago I recorded my first podcast. It's called "Action Potential" (it can be found in iTunes).

There are many tips I learned over the past year. The first step you'll want to take is to practice recording a brief segment of yourself talking on your phone. Then play it back to yourself over and over. Many people don't even realize how they sound until they truly listen to themselves. There may be distracting verbal habits that you do that convey a different attitude than what you would like to express. Once you have an idea of what you sound like it's time to get to work.

Tip #1: Stand or Move When You Are Speaking. I've found that moving while speaking can be very effective. Your voice can shake when you're nervous. By staying in motion you feel more in control and more confident. This also engages your pre-frontal cortex, the part of your brain that is involved in abstract thinking and creative planning. When you are in motion, your ideas and thoughts should flow more smoothly.

A friend of mine interviewed me live on his radio show several months ago. While he had given me an idea of what topics we would discuss, I didn't know the specific questions he would ask. I did everything we've discussed to get into the most powerful mindset before the interview began, but I knew that when we went live there would be elements that I could not control.

For the entire 60-minute show I remained in motion. As I moved, my words flowed well, and my excitement for the topics we discussed was conveyed by the energy in my voice. I've received a lot of positive feedback on the interview, and you can find a copy of it on my website (www.doctorsabrinasegal.com). When you listen to it, you'll hear how controlled my voice sounds. I attribute that mostly to my constant movement throughout the interview. I encourage you to try it.

<u>Tip #2: Use Your Chest Voice and Speak Loudly.</u> Your chest is where all your vocal power comes from. In the same way that we practiced those power poses, try speaking from your chest, not your head. This is similar to the way that people refer to singing. If you use your "head voice," you sing in a falsetto style and there is not a lot of power behind it. If you use your "chest voice," you can belt out the notes in ways that convey strength, purpose, and passion. Singers who use this technique breathe

deeply from their diaphragm. Use this approach to have a strong, confident voice.

Always project louder than you think you should. You'd be surprised at how far you have to go for anyone to think you are being too loud. Ladies, here is an additional tip: Watch the pitch of your voice. When we get excited our voices tend to climb higher in pitch. While it is important to convey excitement in your voice, you want to avoid sounding like a little girl and making it difficult to understand what you are saying. Get into the habit of trying to lower your voice, and let your excitement shine through, by altering the pitch of your voice briefly, rather than allowing it to increase higher and higher. Men don't tend to have this problem as much, because their voices are naturally deeper than women...but men: it doesn't hurt to monitor your pitch as well.

Tip #3: Speak Slowly and Add Purposeful Pauses. When I get excited I tend to talk way too fast, as do most people. When you decrease the speed of your voice, the audience tends to focus more on what you are saying. Some people fear that when they pause to collect their thoughts it makes it seem like they don't know what they are talking about, or that they're not that prepared. Meanwhile, your neurons are firing in milliseconds and helping you gather your upcoming words so that when you do speak, you have a much more articulate way of communicating what you were planning to say before that pause. So allow yourself to take some deep breaths, pause and collect your thoughts.

A purposeful pause can add a dramatic tone to your message, especially if it was intended to allow your audience to think about a point you just addressed, and not because you are searching for the next words you plan to say. When I was watching Bo Eason speak at that seminar, he did something that was incredibly powerful, and something that I have not seen often, even from the most polished speakers.

In the middle of his speech, when he was sharing his story and building an emotional experience for his audience he said, "Imagine a world where people took one step toward their dreams." He paused for about ten seconds and just stared at his audience. When you could hear a pin drop in the room from the silence he said, "That's your job."

It has been several years since I heard that speech and that statement is still with me today. Why? By pausing, he peaked his audience's curiosity, and increased engagement. Sometimes when you pause, you allow your audience to recognize the power of the words you just spoke, or are about to speak. By doing this they will most likely become better listeners, and become much more focused on your message.

Tip #4: Polishing the Power of Your Voice. There are several things you can do to take your vocals to the next level, if you really want to make an impact with your voice. The first thing to do is to notice any nervous verbal habits you have, like saying "um" and try to pause instead, or say something different, that flows with what you are about to mention next. If you notice that your voice sounds dry or scratchy make sure that you always have water nearby, and take a sip every time you pause. If you are in an area where there is a lot of background noise or there is an echo, you can purchase inexpensive microphones from Amazon that connect to your smartphone when you record. These can make your voice sound crystal clear.

Finally, if you have a mentor with a voice you really admire, you can practice enhancing certain aspects of your own voice to reflect the qualities you like in their voice. For example, I adore Amy Porterfield's voice. Her voice is so melodic it almost sounds like you are listening to a Disney Movie, yet Amy is an intelligent and successful business entrepreneur and social media expert. I loved that smooth quality of her voice so much that I started thinking of the way it sounds in my head each time I was recording. Naturally, my voice sounded a bit more melodic as I was talking and keeping the sound of her voice in my head. This tip can be very effective in polishing your own voice.

If you have prepared your notes ahead of time be very careful not to sound like you are reading. Pretend that you are simply having a conversation with someone and you are excited to share this information with them. You can polish the power of your voice by allowing for various vocal tones to reflect different emotions that you are expressing. Vary your pitch briefly so that your audience's ears are constantly stimulated. When you do this, it is difficult for them to stop listening or feel bored, because there is an element of unpredictability that subconsciously keeps them engaged and wanting to hear more. When people sound monotone, not only can it cause listeners to daydream, but it could cause someone who's listening while driving to fall asleep at the wheel!

When you are giving a webinar or teaching an online course, you want your vocal inflections to match the visual representation of the material that you are presenting. For example, if you have an animation on the screen, you want the pace of your voice to match the speed at which different components are moving on the screen. This will capture and intensify the excitement you are expressing, increase engagement, and convey a smooth, powerful and confident message to your audience.

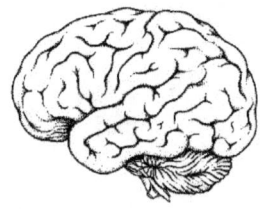

"I want to put a dent in the universe"—Steve Jobs

## Chapter 12: The Power in You

This is it. The moment you've been waiting for. You might not be at this point in your career yet, but I want you to imagine this for a minute. You've been asked to share your message in front of thousands of people at an event that will be broadcast live all over the world. Some of the people you've admired most in your life will be there.

As you jump out of bed, put on your suit, and look at yourself in the mirror, those butterflies flutter in the pit of your stomach. This might be the greatest moment in your career so far. You get your power play list and you start dancing around in your bedroom to those songs that inspire you and lift you up. You really get into it, enjoying yourself and almost laughing with excitement. For a split second, a trace of fear washes over you, reminiscent of past doubts. You shake it off and keep dancing.

While the music plays you allow visual imagery of all the leaders and powerful people you've admired flood your mind. You feel excited that you will join them by accomplishing something that means a great deal to you in such a way that you will be proud. You stand up tall and practice your power poses, holding them for two minutes each. You feel strong, relaxed, and confident. There is no way Mr. Cortisol is going to attend this event. In fact, he is napping right now. Fast asleep. Noradrenaline peaks his head up above the covers every now and then, but he's in a great mood this morning as he is excited and not scared. He too, goes back to sleep.

You drive to the event, the whole time listening to your power play list. When you arrive, a sense of familiarity and peace washes over you, as you have visited the building and the room that you are speaking in many times in preparation. You know where everything is, from the light switches, to the projector. to the seating arrangement. You know it inside and out almost as if it was your own body. And you welcome it as if it belongs to you.

You look at the empty seats, as it is still ten minutes early and people are continuing to trickle into the auditorium. You stand in the back and look up at the stage, imagining yourself up there and picturing what your audience will be thinking and feeling while you are speaking. The excitement builds further. Noradrenaline wants to play! You invite him, but tell him that he needs to sit quietly and focus on your message throughout your presentation. Meanwhile, cortisol is still out like a log, as you remind yourself that they invited you to speak here today because you know more about what you are going to present on this topic than anyone else in the room. There's no one else who could deliver your message with the unique experience and perspective that you bring today.

You step up on that high stage, and look down at your audience. The seats are beginning to fill in more. You start to set up your computer and you get your mug and set it down on the podium. You place a few flash cards on the podium with notes incase you need to glance at them, but you feel confident that you will not need them. Once more, you stand tall, feet apart, arms wide, head high allowing yourself to embrace that power stance without it seeming inappropriate or out of place. For a few moments you picture that you are a lion standing strong, with a predatory stance on that stage. You haven't opened your mouth yet, but there's something about your body language that makes your audience already unable to turn away from you.

The lights dim, and the stage lights are turned up. Suddenly you are standing under those bright lights and everyone is staring at you. Deep breath, head still high with confidence. Thirty seconds to go time. You take a sip from your mug and set it down. And it's the last time your fingertips will brush that podium for the remainder of your talk. You turn to your right and see the camera crew signal that you are going live in 3, 2, 1....and as you open your mouth to speak you take large, powerful steps toward your audience greeting them with your energy, excitement, and passion for your message. As you scan their faces, your gaze connects with one member in the front row whose eyes are full of sparkle and curiosity. There are moments where you turn back to this face throughout your talk, and it strengthens your feelings of confidence and your excitement to share your message.

For the next 45 minutes the audience cannot take their eyes off of you.

Allow that to sink in. What was once your greatest fear has now become one of your unique strengths. You have captured them. While sharing your message and teaching your audience, you remain authentic to who you are. You are no longer trying to "fake it till you make it". You have embraced what is valuable and unique about you and your message and you remain focused on that throughout your presentation. It keeps you centered and it keeps Mr. Cortisol from interfering.

Toward the end of your presentation, in those last few minutes Noradrenaline is skipping around wanting badly to celebrate with you. The last words leap from your mouth and the crowd cheers. The lights are still shining brightly on you, your spotlight. As you stare past the brightness into the sea of people they are rising to their feet. The sound of clapping hands is almost deafening. You stop for a moment and take everything in, your heart starts to race, and you think to yourself, "Okay, it's time to let those stress hormones do their work!" It's time to make some powerful memories, and you allow them to perform that cascade of molecular events inside your brain so that you can remember this moment forever. And with that, you drop the mic... and proudly walk off stage to the sound of cheering. Now, it's time to celebrate!

I want you to know what it's like to experience this success and not be afraid of what could be some of the greatest experiences in your life. If you are ready to do the work, I think the scene I just described to you is not that far out of reach. Keep reaching, learning and taking risks. It's the only way you'll conquer your fears. As Steve Jobs used to say "Stay hungry. Stay Foolish."

### Thank you!

There are many books on the topic of public speaking out there, and you did not have to purchase this book. But you did. My hope is that you were looking for a whole new approach to personal development, where a neurobiologist's personal perspective offered you valuable information that you could not find elsewhere. If I did my job, I left you satisfied with what you have read, and excited to do the work to overcome your fear of public speaking or to strengthen your skills. If you have enjoyed what you read please consider leaving an honest review on Amazon. This will allow the book to reach more people and help make a difference in their lives.

If you have specific feedback about the book please email dr.sabrinasegal@gmail.com with the subject line "Panic to Power book feedback".

It was one of my greatest joys to write this book, to share my personal perspective about a topic that is dear to my heart. I am excited to share my knowledge of the brain in fun, entertaining ways that make it easy to understand. With this new perspective, and accurate knowledge about how your brain works, I believe you can improve your life in many areas. I encourage you to continue reading my book series on different topics about our every day lives, titled "A Neurobiologist's Personal Perspective" and if there is a topic you would like to see covered in the future please email me at: dr.sabrinasegal@gmail.com and list "book topic" in the subject line.

I also invite you to continue to keep up with this book series by joining the email list on www.doctorsabrinasegal.com. I encourage you to connect with my public Facebook group, subscribe to my YouTube channel at: www.youtube.com/c/DrSabrinaSegal, to follow me on twitter: @DrSabrinaSegal, and to like my Facebook page: Dr. Sabrina Segal. You can also tune into my podcast "Action Potential" in the iTunes store. I look forward to seeing you around!

~Sabrina

# Scientific References By Chapter

**Chapter 2:**

Kirschbaum, C et al. The Trier Social Stress Test—a tool for investigating psychobiological stress responses in a laboratory setting. 1993. *Neuropsychobiology*; 28(1-2): 76-81.

**Chapter 3:**

1. Cahill, LF et al. 1994. Beta-adrenergic activation and memory for emotional events. *Nature*; 371(6499): 702-4.

2. McIntyre, CK, Hatfield, T., and McGaugh, J.L. Amygdala norepinephrine levels after training predict inhibitory avoidance retention performance in rats. 2002. *The European Journal of Neuroscience*; 16(7): 1233-6.

3. Roozendaal, B et al. Glucocorticoids interact with emotion-induced noradrenergic activation in influencing different memory functions. 2006. *Neuroscience*; 138(3): 901-10.

**Chapter 4:**

1. Roozendaal, B et al. Glucocorticoids enhancement requires arousal-induced noradrenergic activation in the basolateral

amygdala. 2006. *Proceedings from the National Academy of Sciences*; 103(17): 6741-6.

**Chapter 5:**

1. Cuddy, AJ et al. 2015. Preparatory power posing affects nonverbal presence and job interview performance. *The Journal of Applied Psychology*; 100(4): 1286-95.

**Chapter 9:**

1. Bogert, B et al. 2016. Hidden sources of joy, fear, and sadness: Explicit versus implicit neural processing of musical emotions. *Neuropsychologia* doi: 10.1016/j.neuropsychologia.2016.07.005
2. Segal, SK et al. 2014. Glucocorticoids interact with noradrenergic activation at encoding to enhance long-term memory for emotional material in women. *Neuroscience*; 277: 267-72.

## Personal References by Chapter

**Chapter 9**

1. Beyoncé: A popular American singer, dancer and performer. Her song, "Run the World" is a song that supports and empowers women.

2. Chalene Johnson: Celebrity fitness trainer, New York Times best-selling author, successful business entrepreneur and social media expert. The fashion show in this chapter was an event held at her annual Camp Do More seminar, focusing on fitness and self-improvement. Her Smart Success Academy is a business seminar for entrepreneurs. You can find my YouTube testimonial at: https://www.youtube.com/watch?v=SFwTf6C7BGM. For more information on Chalene Johnson please visit: www.chalenejohnson.com.

3. Bo Eason, former NFL football player, public speaker, actor, playwright. He wrote, produced, and starred in the play, "Runt of the Litter". It has been referred to as the "most powerful play of the decade." For additional information go to: www.boeason.com

**Chapter 10:**

1. Channel Five News: news station in Arizona. The event was the Science of Baseball held at Scottsdale Baseball field.

2. Live Streaming Platforms: Facebook, Periscope.

**Resources for Public Speaking:**

1. Websites to improve communications skills:

    Ellie Parvin, Communications Expert and Coach:

    www.ellieparvin.com

    Bo Eason, Communications Expert: www.boeason.com

    Chalene Johnson, Communications Expert:

    www.chalenejohnson.com

2. Podcasts:

    Chalene Johnson's "Build Your Tribe" Podcast:

    http://www.chalenejohnson.com/podcast/

Amy Porterfield's "Online Marketing Made Easy" Podcast:

http://www.amyporterfield.com/category/podcast/

3. Webinars and Online Courses:

   Amy Porterfield:

   http://www.amyporterfield.com/webinarsthatconvert/

4. Social Media Experts on Live Streaming:

   Kim Garst: http://kimgarst.com/

   Mari Smith: http://www.marismith.com/